Slithering Snakes
BLACK MAMBA

FAST AND DEADLY

BY NATALIE HUMPHREY

Please visit our website, www.enslow.com. For a free color catalog of all our high-quality books, call toll free 1-800-398-2504 or fax 1-877-980-4454.

Library of Congress Cataloging-in-Publication Data

Names: Humphrey, Natalie, author.
Title: Black mamba : fast and deadly / Natalie Humphrey.
Description: New York : Enslow Publishing, [2021] | Series: Slithering snakes | Includes index.
Identifiers: LCCN 2019057130 | ISBN 9781978517691 (library binding) | ISBN 9781978517677 (paperback) | ISBN 9781978517684 (6 Pack) | ISBN 9781978517707 (ebook)
Subjects: LCSH: Black mamba—Juvenile literature.
Classification: LCC QL666.O64 H86 2021 | DDC 597.96/42—dc23
LC record available at https://lccn.loc.gov/2019057130.loc.gov/2019060176 2019060212

Published in 2021 by
Enslow Publishing
101 West 23rd Street, Suite #240
New York, NY 10011

Copyright © 2021 Enslow Publishing

Designer: Sarah Liddell
Editor: Natalie Humphrey

Photo credits: Cover, pp. 1 (black mamba), 5, 8–9, 19 NickEvansKZN/Shutterstock.com; background pattern used throughout Ksusha Dusmikeeva/Shutterstock.com; background texture used throughout Lukasz Szwaj/Shutterstock.com; p. 7 1001slide/iStock/Getty Images Plus/Getty Images; p. 11 © iStockphoto.com/Robert Styppa; p. 13 Adre Coertzer/Shutterstock.com; p. 15 suebg1 photography/Moment/Getty Images; p. 17 Heiko Kiera/Shutterstock.com; p. 21 Anabela88/Shutterstock.com.

Portions of this work were originally authored by Angelo Gangemi and published as *Black Mambas*. All new material this edition authored by Natalie Humphrey.

All rights reserved. No part of this book may be reproduced in any form without permission in writing from the publisher, except by a reviewer.

Printed in the United States of America

Some of the images in this book illustrate individuals who are models. The depictions do not imply actual situations or events.

CPSIA compliance information: Batch #BS20ENS: For further information contact Enslow Publishing, New York, New York, at 1-800-398-2504.

CONTENTS

Fast and Deadly 4
At Home on the Savanna 6
Keep Far Away! 12
A Deadly Bite 16
Baby Mambas 18
Black Mambas and People . . 20
Words to Know 22
For More Information 23
Index 24

Boldface words appear in Words to Know.

FAST AND DEADLY

The black mamba is one of the fastest snakes around—and one of the deadliest! When this snake is cornered or angry, it **strikes** back. Its bite has deadly **venom** that is dangerous to humans. If you ever see a black mamba, stay back!

BLACK MAMBA

AT HOME ON THE SAVANNA

The black mamba lives on the African **savanna** and in rocky forests. Black mambas like hot, dry, and rocky places. During the day, they **slither** and hunt for **prey**. At night, they sleep under rocks, in holes, and even in trees!

THE BLACK MAMBA IS AFRICA'S LONGEST VENOMOUS SNAKE.

7

Black mambas have brown, gray, or greenish scales with white stomachs. Their name comes from the black color inside of the snake's mouth. When a black mamba is scared or angry, it opens its mouth wide to try and scare the danger away.

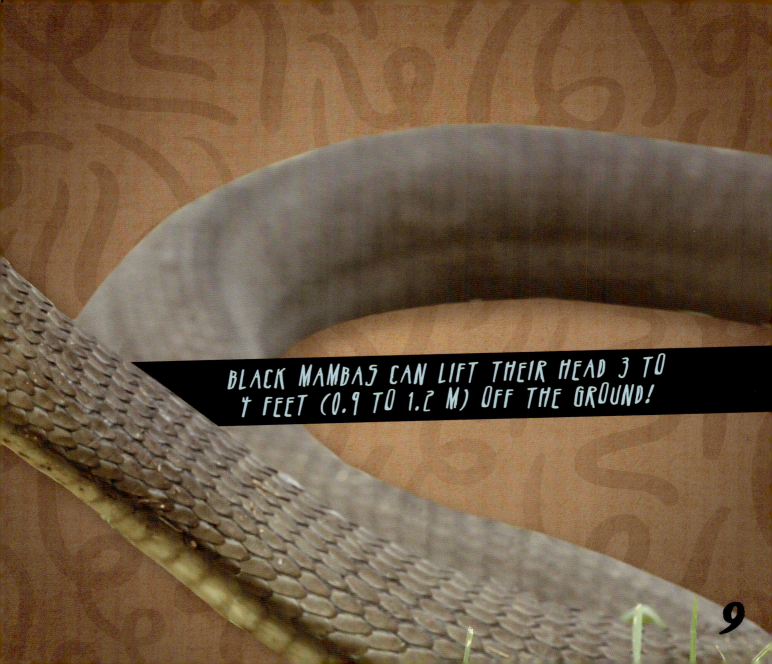

Black mambas can lift their head 3 to 4 feet (0.9 to 1.2 m) off the ground!

Black mambas can be anywhere between 6 feet (1.8 m) and 14 feet (4.3 m) long. They weigh around 3.5 pounds (1.6 kg). The black mamba is one of the fastest snakes in the world. Black mambas can move faster than 12 miles (19 km) per hour!

A BLACK MAMBA'S BOX-SHAPED HEAD HELPS IT MOVE FASTER.

KEEP FAR AWAY!

Black mambas are shy snakes and would rather run from danger than fight. But if it is cornered, the black mamba will lift its head and open its mouth. It hisses and makes it neck wider so it looks scarier.

Black mambas are in the same family as cobras.

When a black mamba attacks, it bites its prey several times. Its **fangs** shoot venom into the prey's body. The venom is so strong that it can even kill a lion! If a black mamba bites a person, the person must take **medicine** right away.

A DEADLY BITE

Black mambas hunt on the ground and in trees. Sometimes they eat birds, but they usually eat mice, rats, and squirrels. When a black mamba finds its meal, it bites the prey and waits for the venom to kill it. Then, it swallows its dinner whole!

Black mamba venom can kill prey in less than 20 minutes!

Baby Mambas

Female black mambas lay between 6 and 17 eggs every year. In about three months, the newborn black mambas break through their eggs. Baby black mambas are 1.3 to 2 feet (40 to 60 cm) long when they are born. They are ready to start hunting right away!

Baby mambas use an **egg tooth** to break through their egg.

BLACK MAMBAS AND PEOPLE

Black mambas don't like being around people. Sometimes, they accidentally make their home on an African farm. When they see people nearby, the black mamba will try to escape, but people should still be careful. One bite from a black mamba can be deadly!

WORDS TO KNOW

egg tooth A hard, toothlike bump used to help young snakes when they're hatching, or coming out of their egg.

fang A long, sharp tooth.

medicine Matter that is used to treat sickness or help with pain, usually given by a doctor.

prey An animal that is hunted or killed by another animal for food.

savanna A large, flat area of land with grass and very few trees in Africa and South America.

slither To slide easily over the ground.

strike To hit something in a forceful way.

venom Something a snake makes in its body that can harm other animals.

FOR MORE INFORMATION

BOOKS

Kingsley, Imogen. *Black Mambas*. Minneapolis, MN: Jump!, Inc., 2017.

Pope, Kristen. *On the Hunt with Black Mambas*. Mankato, MN: The Child's World, 2016.

WEBSITES

African Snakebite Institute
www.africansnakebiteinstitute.com/snake/african-snakes-black-mamba/
Watch videos and check out more photos of black mambas!

National Geographic
www.nationalgeographic.com/animals/reptiles/b/black-mamba/
Discover more photographs of black mambas in the wild.

Publisher's note to educators and parents: Our editors have carefully reviewed these websites to ensure that they are suitable for students. Many websites change frequently, however, and we cannot guarantee that a site's future contents will continue to meet our high standards of quality and educational value. Be advised that students should be closely supervised whenever they access the internet.

INDEX

baby mambas, 18

color, 8

fangs, 14

food, 16

home/where it lives, 6, 20

hunting, 6, 16, 18

length, 10

name, 8

prey, 6, 14, 16

savanna, 6

speed, 4, 10

trees, 6, 16

venom, 4, 14, 16

weight, 10